20 THINGS YOU DIDN'T KNOW ABOUT WEATHER

DOUG BRADLEY

PowerKiDS press

Published in 2023 by The Rosen Publishing Group, Inc.
2544 Clinton Street, Buffalo, NY 14224

Portions of this work were originally authored by Caitie McAneney and published as *20 Fun Facts About Weather*.
All new material in this edition was authored by Doug Bradley.

Editor: Greg Roza
Book Design: Tanya Dellaccio

Photo Credits: Cover Drew McArthur/Shutterstock.com; p. 5 54115341/Shutterstock.com; p. 6 Samrat Sengupta/Shutterstock.com; p. 7 sirtravelalot/Shutterstock.com; p. 8 https://upload.wikimedia.org/wikipedia/commons/f/fc/Hankow_city_hall.jpg; p. 9 Abhilekh Saikia/Shutterstock.com; pp. 10, 21 Everett Collection/Shutterstock.com; p. 11 Brian Nolan/Shutterstock.com; p. 12 MISHELLA/Shutterstock.com; p. 13 Captainz/Shutterstock.com; p. 14 V_Sot_Visual_Content/Shutterstock.com; p. 15 CSNafzger/Shutterstock.com; p. 16 Plume Photography/Shutterstock.com; p. 17 https://upload.wikimedia.org/wikipedia/commons/7/79/1993_storm_century.jpg; p. 18 SkyLynx/Shutterstock.com; p. 19 https://upload.wikimedia.org/wikipedia/commons/a/aa/Olivia_Apr_10_1996_1123Z.png; p. 20 John D Sirlin/Shutterstock.com; p. 22 Rainer Lesniewski/Shutterstock.com; p. 23 ellepistock/Shutterstock.com; p. 24 Pau Buera/Shutterstock.com; p. 25 (main) Jana Janina/Shutterstock.com; p. 25 (inset) Wirestock Creators/Shutterstock.com; p. 26 Christian Wilkinson/Shutterstock.com; p. 29 Gorodenkoff/Shutterstock.com.

Cataloging-in-Publication Data
Names: Bradley, Doug.
Title: 20 Things You Didn't Know About Weather / Doug Bradley.
Description: New York : PowerKids Press, 2023. | Series: Did You Know? Earth Science| Includes glossary and index.
Identifiers: ISBN 9781538389805 (pbk.) | ISBN 9781538389829 (library bound) | ISBN 9781538389836 (ebook)
Subjects: LCSH: Climatic Changes–Juvenile literature | Weather–Juvenile literature.
Classification: LCC QC981.8.C53 B73 2023 | DDC 551.55–dc23

Manufactured in the United States of America

CPSIA Compliance Information: Batch #CWPK23. For Further Information contact Rosen Publishing at 1-800-237-9932.

CONTENTS

WILD WEATHER

Earth has many different kinds of weather. Some areas are sunny most of the year. Others feature lots of snow, rain, and wind. Sometimes it's very hot, and sometimes it's so cold you can't go outside! A long-term pattern of weather is called a **climate**. The climate is different depending where on Earth you are.

The weather can also be dangerous. People can't control the forces of nature. However, scientists are constantly learning more about climate and weather on Earth.

The more we know about Earth's weather patterns the easier it will be to prepare for dangerous weather phenomenon.

RAIN EXTREMES

DID YOU KNOW?

Mawsynram, India, is most likely the rainiest place on Earth.

The village of Mawsynram gets between 450 inches and 700 inches (1,140 and 1,780 cm) of rain a year! During the rainy season—from June to September—the village only receives about five hours of sunshine a month.

Mawsynram's rainy season is caused when winds called monsoons bring warm, moist weather to the area once a year.

We usually picture hot, sandy places when we think of deserts. But there are also cold deserts! There are four kinds of deserts, and three of them have cold or cool climates.

RECORD FLOODS

DID YOU KNOW?

The 1931 floods in China affected about one-fourth of China's population at the time.

China's great rivers overflowed when they became filled with huge amounts of melting snow and pouring rain. Over 51 million people lost their homes, and as many as 3.7 million may have died.

On August 19 in the city of Hankow, shown here, the floodwaters rose to a high-water mark of 53 feet (16 m) higher than average.

India is famous for its monsoon season, which often includes flooding. When floodwaters rose to more than 166 feet (50.6 m) in some places, they broke flooding records!

HURRICANES!

DID YOU KNOW?

The Galveston hurricane of 1900 created a wave that washed over Galveston Island and the city of Galveston, Texas, itself.

The hurricane's winds likely reached speeds of more than 140 miles (193 km) per hour. The **storm surge** raised the sea level 15.7 feet (4.8 m)! The highest **elevation** on Galveston Island at the time was 8.7 feet (2.7 m).

When the Galveston hurricane hit the Texas coast on September 8, 1900, there was little warning. In fact, the warning system used in Galveston at that time was a man in a horse-drawn cart!

Hurricane Katrina caused widespread damage. Many people lost everything, including homes, cars, boats, and pets.

DID YOU KNOW?

Wind speed during Hurricane Katrina surpassed 170 miles (274 km) per hour!

Hurricane Katrina hammered Florida, Mississippi, Alabama, Georgia, and Louisiana in 2005. The hardest-hit area was New Orleans. Winds and waves broke down **levees** and the city became flooded.

DID YOU KNOW?

In October 2012, Hurricane Sandy hit the Caribbean. It was so huge that people called it "Superstorm Sandy" and "Frankenstorm"!

Hurricane Sandy caused great damage in countries like Haiti. Then it traveled up the U.S. East Coast, reaching the New Jersey shore and New York City. Hurricane Sandy destroyed an amusement park on the Jersey shore.

During Hurricane Sandy, a wave in New York Harbor was measured at 32.5 feet (9.9 m)! The New York City subway system flooded.

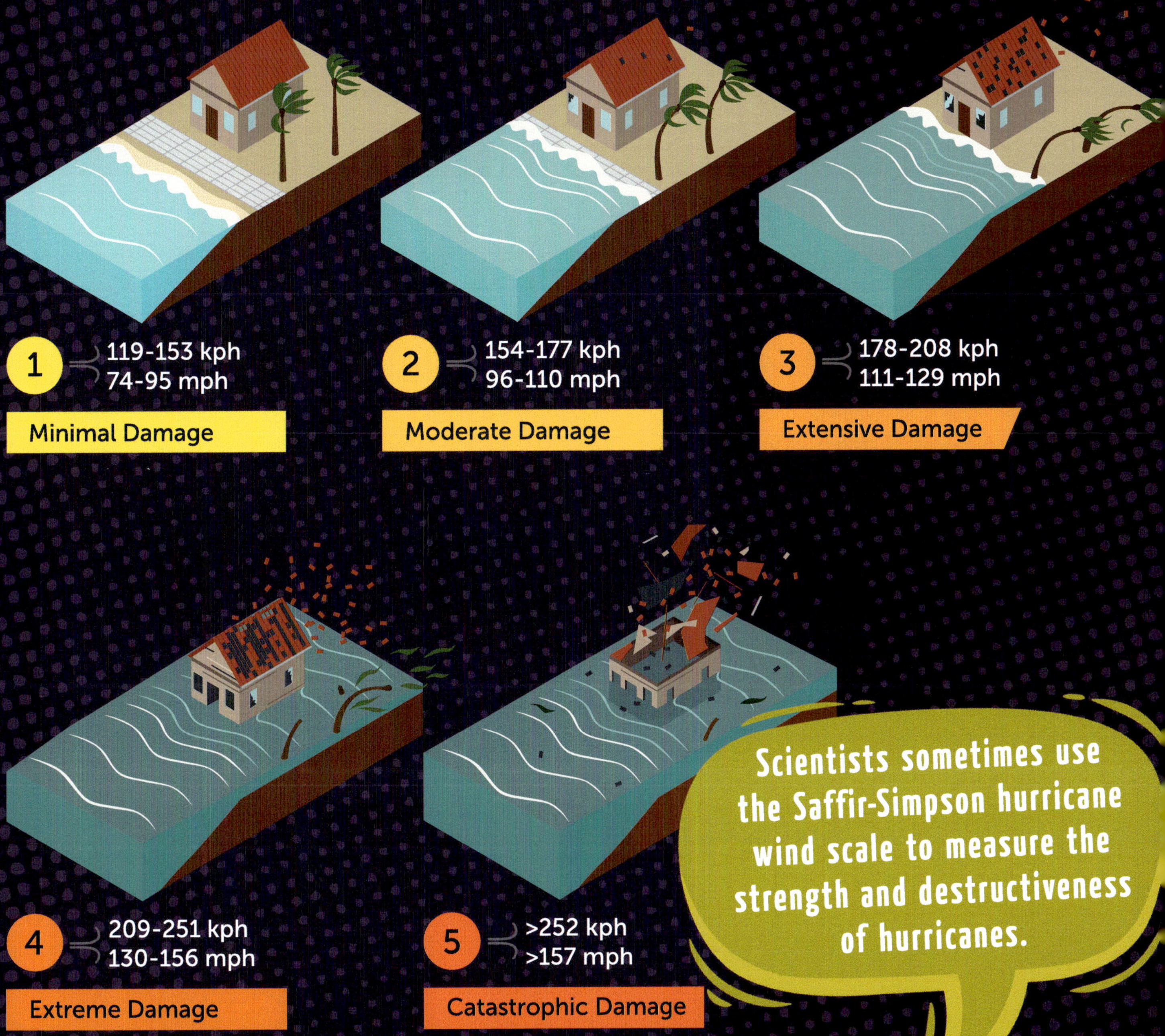

1 119-153 kph
74-95 mph
Minimal Damage

2 154-177 kph
96-110 mph
Moderate Damage

3 178-208 kph
111-129 mph
Extensive Damage

4 209-251 kph
130-156 mph
Extreme Damage

5 >252 kph
>157 mph
Catastrophic Damage

Scientists sometimes use the Saffir-Simpson hurricane wind scale to measure the strength and destructiveness of hurricanes.

POWERFUL BLIZZARDS

DID YOU KNOW?

The word "blizzard" once meant "a violent blow." It was first used to describe snowstorms in the United States around 1859.

A blizzard is a storm with blowing snow and low **visibility**. When winds that are more than 35 miles (56 km) an hour blow the snow around, no one can see!

If drivers can't see the road because of blowing snow, it might be considered a blizzard, even if there's not much snow falling.

DID YOU KNOW?

In 1921, a monster snowstorm hit Silver Lake, Colorado, with 75.8 inches (192.5 cm) of snow in one day!

Silver Lake holds the U.S. record for the most snowfall in 24 hours. Over 32.5 hours, Silver Lake received a total of 95 inches (241 cm). That's taller than a tall adult!

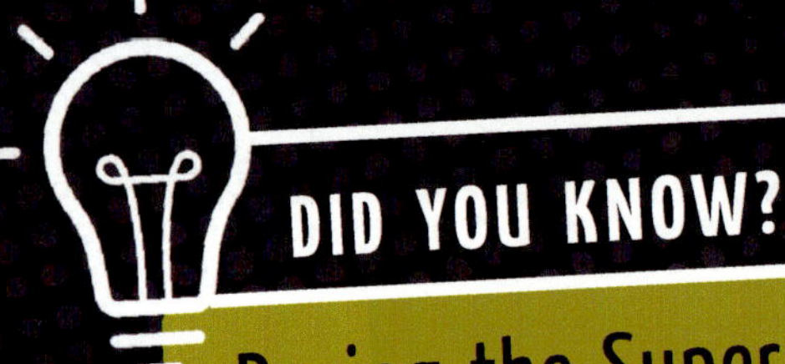

DID YOU KNOW?

During the Super Bowl Blizzard of 1975, wind blew snow into drifts that reached over 20 feet (6 m) tall!

Strong winds can create large slopes of snow called snowdrifts. During the Great Blizzard of 1975, also known as the Super Bowl Blizzard, several southern and midwestern states experienced the tallest snowdrifts they had ever seen!

Minnesota was hit particularly hard during the 1975 blizzard. The Minnesota Vikings lost the 1975 Super Bowl (held in New Orleans, Louisiana) to the Pittsburgh Steelers by a score of 16-6.

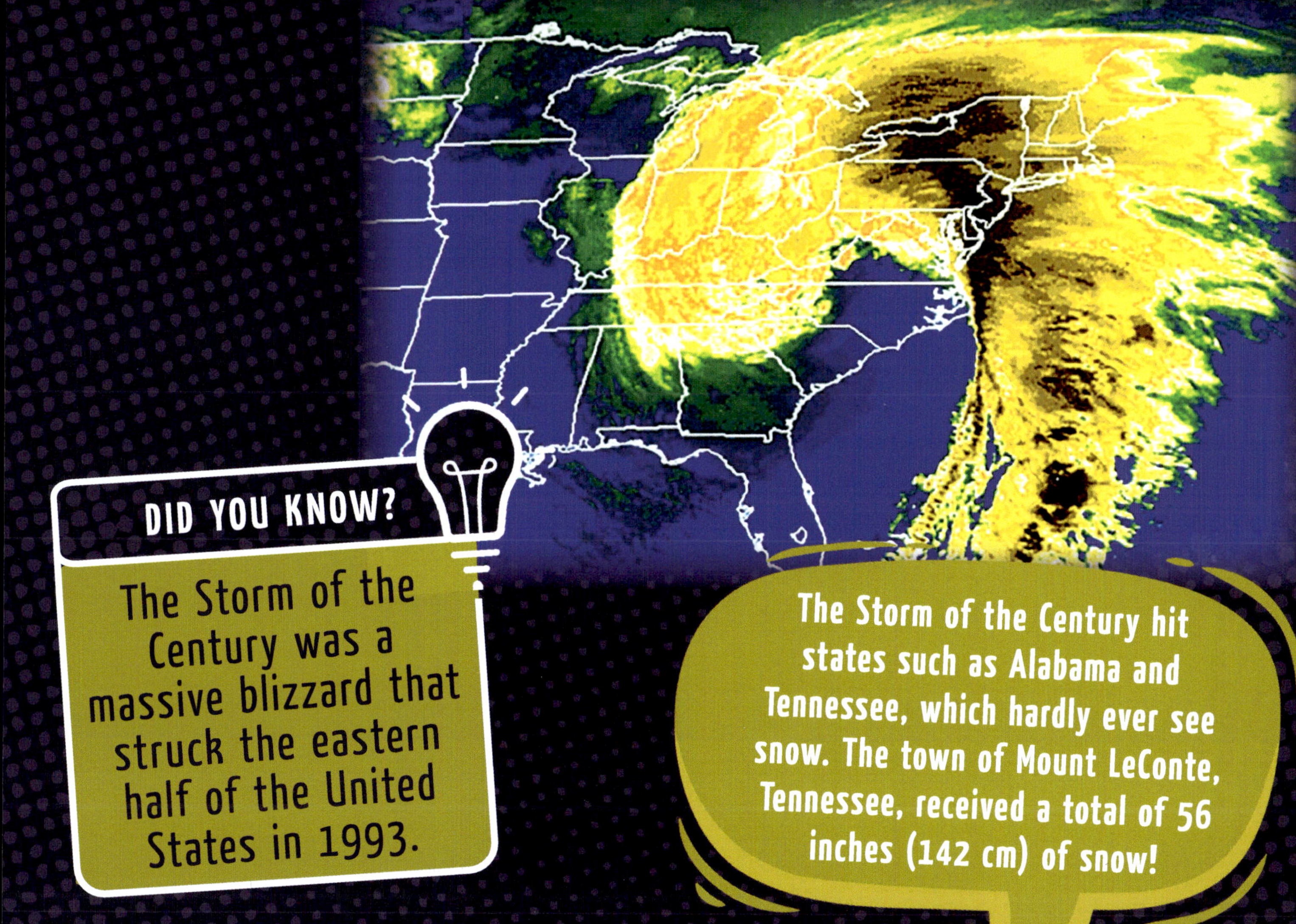

DID YOU KNOW?

The Storm of the Century was a massive blizzard that struck the eastern half of the United States in 1993.

The Storm of the Century hit states such as Alabama and Tennessee, which hardly ever see snow. The town of Mount LeConte, Tennessee, received a total of 56 inches (142 cm) of snow!

A **cyclone** that formed over the Gulf of Mexico brought tons of snow from the Southeast to the Northeast United States. It caused approximately $5.5 billion in damages!

WILD WINDS

DID YOU KNOW?

People have been using wind power for thousands of years—and still do!

People first used wind to power sailboats more than 5,000 years ago. Today, we use wind to make electricity. Wind is a renewable resource.

Wind turbines are giant machines that turn wind into electricity.

During Cyclone Olivia in 1996, Barrow Island, Australia, was hit by a gust traveling 253 miles (407 km) per hour! A gust is a burst of wind that lasts for just a few seconds.

DUST STORMS

DID YOU KNOW?

"Haboob" is an Arabic word for walls of sand caused by strong winds.

Dust storms happen when powerful winds blow over a sandy desert. They can sweep over towns and cities and bury them in sand. Dust storms can reduce visibility to zero, making driving dangerous.

Haboobs mostly happen in the southern Sahara Desert, but they have been spotted in the Middle East, North America, Australia, and the planet Mars!

In the 1930s, workers cleared large areas of land in the Great Plains for wheat farming. The lack of tree roots loosened the soil, causing it to blow away during high winds. Frequent dust storms covered a large area across several states.

DID YOU KNOW?

Scientists estimate that the Great Plains lost 1.2 billion tons (1.09 mt) of soil in just one year during the period that became known as the Dust Bowl.

TORNADO TIME

DID YOU KNOW?

The United States has more tornadoes—around 1,200 per year—than any other country in the world!

Tornadoes often happen as part of huge thunderstorms called supercells. Part of the United States is known as Tornado Alley because so many tornadoes form there!

States that are part of Tornado Alley include Oklahoma, Texas, Kansas, and Nebraska.

DID YOU KNOW?

Waterspouts sometimes cause frogs and fish to rain from the sky!

Waterspouts form when air spins quickly over open water. They can pick up animals while they're over water. The water and animals drop back down to Earth from clouds in the sky.

A FLASH OF LIGHT

DID YOU KNOW?

A bolt of lightning can make the air around it five times hotter than the surface of the sun!

Lightning is a huge spark of electricity that forms between clouds and the ground. Lightning can heat or burn anything in its way. The air around lightning reaches 54,000 degrees Fahrenheit (30,000 degrees Celsius)!

Lightning can sometimes strike over 10 miles (16 km) from the storm cloud where it started.

This happens because the heat quickly turns the water and sap inside the tree to **vapor**. This creates steam, which blows the tree apart when it escapes!

EARTH'S CHANGING CLIMATE

Earth's weather and climate are going through rapid changes, partly due to human activities.

Climate change, which is a change in weather patterns over time, is being sped up by human activities, such as burning fossil fuels. This rapid climate change is causing dangerous storms to happen more often.

Rising temperatures on Earth are causing glaciers and ice sheets in Antarctica to melt. If this continues, global sea levels will rise and put coastal cities in danger.

GLACIERS AND ICE SHEETS SHRINK

SOME PLACES WILL GET COOLER

SEA LEVEL RISES

MORE HARMFUL WEATHER EVENTS

SOME PLACES WILL GET HOTTER

ANIMALS AND PLANTS DIE OFF

PREPARED FOR THE WEATHER

Meteorologists are scientists who study weather. They do their best to **predict** future weather, but Earth's weather can be unpredictable.

Because of climate change, bad weather may happen more often in the future. Be prepared! Pack important supplies, such as food, water, clothing, and a radio. You may also need a can opener, flashlights, batteries, tools, water, and a first aid kit. Be prepared to head to a safe place if you're told to do so.

Once in a safe spot, pay attention to TV or radio news to know when bad weather has passed!

GLOSSARY

climate: The average weather conditions of a place over a period of years.

cyclone: Another word for a hurricane.

elevation: Height above sea level.

levee: A long wall of soil built along a coast or river to prevent flooding.

phenomenon: A fact or event that can be observed.

predict: To guess what will happen in the future based on facts or knowledge.

storm surge: A rise in sea level during a storm.

submerged: Underwater.

vapor: Matter in the form of a gas or very small drops mixed with air.

visibility: How far you are able to see because of weather or darkness.

FOR MORE INFORMATION

BOOKS

DK Eyewitness. *Natural Disasters.* New York, NY: DK Children, 2022.

DK Eyewitness. *Weather.* New York, NY: DK Children, 2022.

Farndon, John, Sean Callery and Miranda Smith. *Weather.* New York, NY: Scholastic Nonfiction, 2020.

WEBSITES

Climate Kids
climatekids.nasa.gov
Find out much more about weather from this colorful and detailed website.

Weather Wiz Kids
www.weatherwizkids.com
Learn more about weather and predicting weather from meteorologist Crystal Wicker.

Publisher's note to educators and parents: Our editors have carefully reviewed these websites to ensure that they are suitable for students. Many websites change frequently, however, and we cannot guarantee that a site's future contents will continue to meet our high standards of quality and educational value. Be advised that students should be closely supervised whenever they access the internet.

INDEX